AMAZING WORLD FACTS

Jenny Wood and Anne Civardi
Illustrated by Matthew Cook and John Raynes
Designed by Jane Warring
Series Editor: Heather Amery
Art Editor: Rowena Alsey

Contents

MITCHELL BEAZLEY

Around the World

The world is full of amazing places and natural wonders, from burning hot deserts and the icy cold Poles to steamy forests and towering mountains. The millions of people who live on the seven continents speak many different languages, and have different ways of life, customs and faiths.

The Earth is about 4,600 million years old and weighs about 6,600 million, million, million tonnes. It is about 12,756 km (7,926 miles) across at the Equator and about 40,075 km (24,901 miles) all the way around. The Earth is not completely round but is slightly flattened at the top and bottom.

The Earth rotates about once every 23 hours, 56 minutes and 4 seconds. It revolves around the Sun every 365 days, 6 hours, 9 minutes and 10 seconds.

Inside the Earth

The Earth's surface covers about 510 million sq km (197 million sq miles) and is made up of several layers. The outer shell, called the crust, is cold rock. Under it is the mantle, a layer of hot rock. This surrounds an outer core of liquid metal.

The centre of the Earth, the inner core, is probably a big ball of solid metal with a temperature of about 5,000°C (9,000°F).

The Earth's crust is made up of gigantic pieces of rock called plates, some as thick as 64 km (40 miles). Some of the plates carry the continents, others lie under the ocean floors.

The plates are always moving, very slowly reshaping the Earth's crust. This movement causes volcanoes and earthquakes.

Scientists believe that the continents once formed a huge land mass, called Pangaea. Over millions of years this mass split apart and the pieces slowly moved away from each other to form the seven continents.

Outside the Earth

A layer of gas, called the atmosphere, surrounds the Earth. It is made up of 78% nitrogen and 21% oxygen. It contains other gases, water vapour and dust.

Without the atmosphere nothing could live on Earth. It provides the air we breathe and protects the Earth from harmful rays from the Sun. The higher you go, the thinner the air.

The atmosphere is made up of four layers. Weather happens in the layer nearest to the Earth, the troposphere. The highest layer is made of helium, oxygen and hydrogen.

Over 5,200 million people live in the world. At the present rate of growth, the Earth's population doubles every 41 years.

About $^{3}/_{5}$ of the people live in Asia and Europe. China has the largest population with over 1,000 million people – over four times as many people as in the USA.

There are about 3,000 different languages spoken on Earth. Nineteen have over 50 million speakers each.

There are thousands of different religions in the world but only eight main ones. These are Christianity, Islam, Buddhism, Confucianism, Hinduism, Shinto, Taoism and Judaism. Christianity is the largest religion with more than 1,000 million followers.

The deepest spot in the ocean is the Mariana Trench in the Pacific Ocean. It is about 11,000 m (36,089 ft) deep.

About 70% of the Earth's surface is covered with water. There are three great oceans – the Pacific, Atlantic and Indian Oceans. The biggest and deepest is the Pacific Ocean which covers almost 30% of the Earth's surface. It is about three times larger than Asia.

Where People Live

All over the world, people live in different kinds of homes. The houses they build often depend on where they live, the way they live and the weather they have. Many people use materials that are easy to find and use or are cheap to buy. Some live in houses made of mud, dung, leaves or reeds, on boat houses or under the ground. Others make their homes under cardboard shelters and even in hollowed out rocks.

Cone-shaped Houses

Thousands of years ago, a volcano erupted in Cappadocia, Turkey. The lava and ash hardened into rock which slowly wore away into cone shapes. People have lived inside the cones for over 1,000 years.

Under the Ground

To protect themselves from the hot desert sun, the Troglodyte Arabs of Tunisia build their houses deep under the ground around a big, sunken courtyard.

Home, Sweet Home

Over 17 million people live in Mexico City, one of the world's most over-crowded cities. Many live in shanty towns in houses made of corrugated iron, scraps of wood or metal and even cardboard.

Did You Know?

No women or children are allowed near the monasteries on Mount Athos in Greece where about 1,000 monks live. Perched on the tops of cliffs, some were built nearly 1,000 years ago.

The Efe pygmies, who live in the Ituri Forest in Zaire, live like prehistoric people. They travel from place to place, hunting for meat and gathering roots, fruits and nuts. The hunter who first wounds an animal is given the biggest piece of meat.

High up in the Himalayas in Tibet, family groups of nomads live in tents made of yak hair. There are no trees on these vast grassy plains and it is too cold to grow any crops. Herds of sheep, goats and yaks provide plenty of milk and meat.

The highest dwellings in the world are in Basisi on the border of India and Tibet. They are 5,980 m (19,784 ft) above sea level.

On the Water

The people who live on the banks of the Mekong River, in Cambodia, build their houses on stilts to protect themselves from the river which floods every year.

In some parts of Asia, there is not enough land for houses. In Hong Kong many people live in house boats. They cook, eat and sleep in the same tiny space.

A Way of Life

There are over 5,200 million people in the world and 220,000 babies are born every day. All these people live in many different ways. The food they eat and the work they do often depends on whether they live high up in the mountains, by the sea, in forests, in crowded cities, or in a very hot or cold place.

Life at the Top

High up in the Andes Mountains, the air has little oxygen. The Indians who live there have bigger lungs than most people so they can breathe more easily.

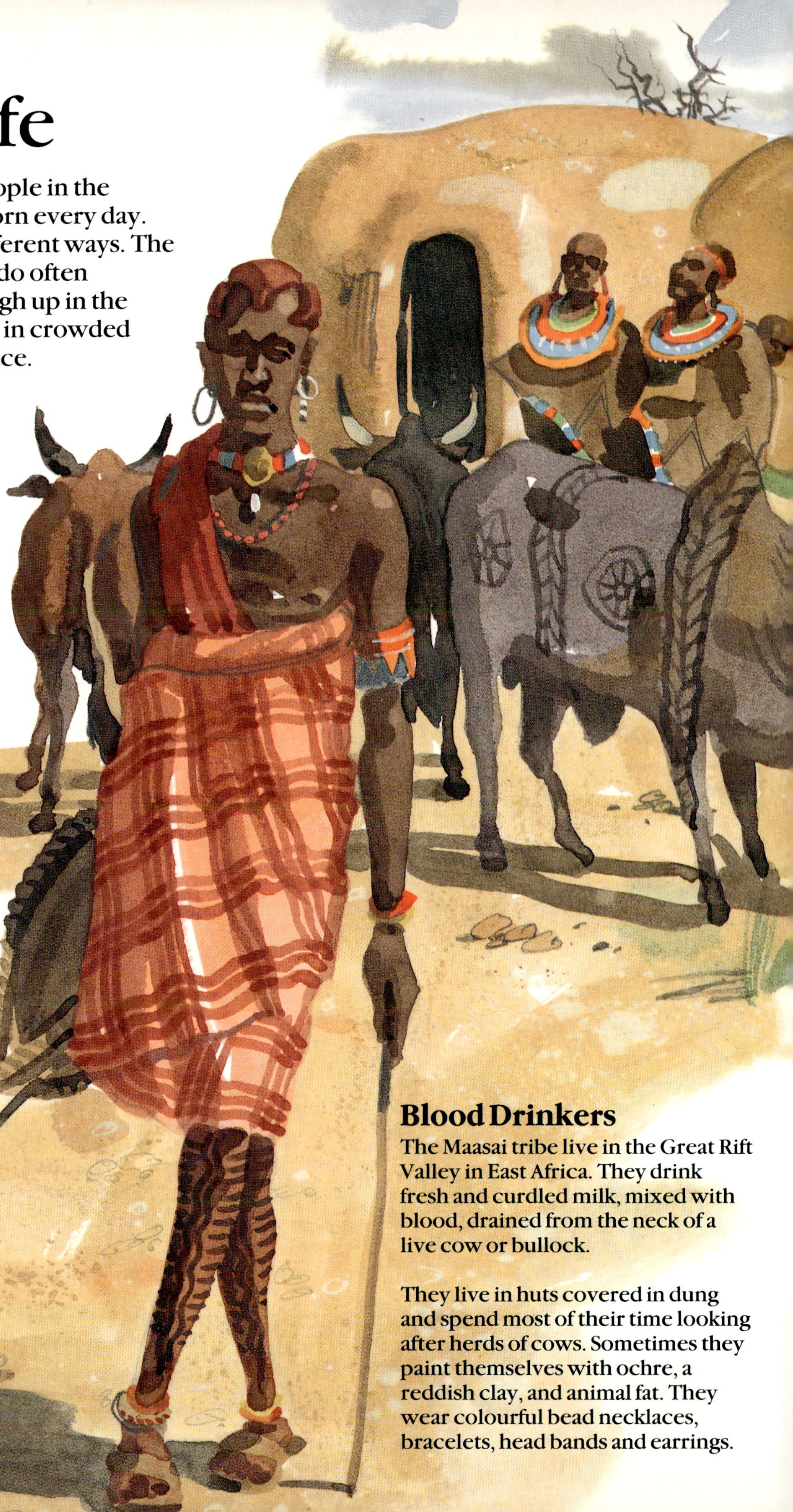

Blood Drinkers

The Maasai tribe live in the Great Rift Valley in East Africa. They drink fresh and curdled milk, mixed with blood, drained from the neck of a live cow or bullock.

They live in huts covered in dung and spend most of their time looking after herds of cows. Sometimes they paint themselves with ochre, a reddish clay, and animal fat. They wear colourful bead necklaces, bracelets, head bands and earrings.

Dying Fast

Thousands of Yanomami Indians, who live in the rainforests in Brazil, are dying early. Miners, searching for gold, have ruined their forests and given them diseases, such as measles.

A Dangerous Job

In South-East Asia, swiftlets make their nests out of their own saliva, high up on cave walls. For hundreds of years, men have risked their lives to collect the nests. They sell them to cooks who make bird's nest soup, a Chinese delicacy.

Blue Men of the Desert

The Tuareg people of North Africa are called "blue men of the desert" because the blue of their robes rubs off on their skin. They dig up roots, called alliums, to eat.

Did You Know?

On the Day of the Dead, Mexicans have picnics in the cemeteries where their families are buried. They eat skulls made of sugar and chocolate coffins as well as bread patterned with skulls and cross-bones.

Every May 5th, Japanese families fly giant fish-shaped kites in honour of their young sons.

The Maori people of New Zealand greet each other by pressing their noses together.

At big feasts, Bedouin Arabs sometimes eat camels stuffed with eggs, chicken and a sheep.

Super Structures

Buildings come in all shapes and sizes, from giant skyscrapers to small mud huts. Made of stone or wood, mud or glass, reeds or concrete, or even cow dung, they are built to live in, to work in or to play in. Architects are always looking for new materials, new designs and new ways of putting up buildings.

The Greatest Wall

The Great Wall winds its way across the hills and mountains of Northern China for nearly 6,400 km (4,000 miles). Work on it began over 2,000 years ago, when Emperor Shih Huang Ti decided to protect his country against invaders.

The Great Wall is the longest structure ever built and was put up completely by hand.

In the east, the sides of the Wall are made from stone or brick and the inside is filled with earth. The top is paved with bricks to make a road.

In the west, the Wall runs through hills and desert. Here the Wall is built of mud bricks. Workmen made them from earth and water and dried them in the sun.

At one time, more than 300,000 men worked for ten years building the huge Wall.

There are watch towers along the length of the wall. They were used by soldiers on guard.

About $6\frac{1}{4}$ Great Walls, laid end to end, would reach all the way around the Equator.

The Great Wall is the only man-made structure which can be seen from the Moon.

Welcome to New York

The Statue of Liberty, which stands in New York Harbour, was first built in Paris. It was shipped to New York in 214 wooden crates. When it was rebuilt, the head and arms were at first put back in the wrong positions.

Did You Know?

The Las Vegas Hilton is one of the world's biggest hotels. It has 3,174 rooms, 13 restaurants and over 3,600 staff. The roof terrace is about the same size as nine or ten soccer pitches.

The Hearst Ranch, in California, is the most expensive private house ever built. Costing $30 million, it has over 100 rooms and a garage big enough for about 25 cars.

A Fairytale Castle

This castle was built over 100 years ago for King Ludwig II of Bavaria. It has watch towers, drawbridges and turrets. Inside there are carvings on every staircase and window, and paintings on every wall.

Keeping Cool

The Hawa Mahal, or "Palace of the Winds", was built in 1799 in the Indian city of Jaipur. It was designed with over 900 windows so that air would blow through the rooms, keeping the ladies of the palace cool.

Inside Out

Brightly-coloured pipes, lifts and escalators are built on the outside of the Pompidou Centre, in Paris, France. Many Parisians would not go near it when it opened in 1977 but it had over 6 million visitors in its first year.

Burning Hot Places

Deserts are the driest places in the world with less than 25 cm (10 ins) of rain a year. They cover 20% of the Earth's land surface. Rocky and bare, only a few are covered in sand and huge sand dunes that have been heaped up by the wind. In the Sahara, the biggest desert in the world, there are mountains as high as 3,300 m (10,827 ft).

Most deserts are very hot in the daytime and cold at night. Few living things can survive in such hot, dry places. Those that do have to find ways of living with little water. Many desert animals never drink at all. Plants store it in their stems and leaves. And people dig underground wells to find water for themselves, their herds and their crops.

Big Drinkers

A camel stores fat in its hump to use when there is no water to drink. After a long time without water, it can drink up to 200 litres (53 gallons) in one go. In winter, it may not drink at all for several months.

Thousands of years ago, the Red Sea covered part of the Danakil Desert. The Earth rose, splitting the sea in two and the water dried up. It left a thick layer of salt in what is now Lake Karum.

Shifting Sands

Strong winds often blow across the desert, piling up the sand into big dunes. Some sand dunes are over 250 m (820 ft) high and more than 100 km (62 miles) long.

The wind blows the sand up one side and down the other. As they creep forward, by as much as 40 m (132 ft) a year, the dunes bury everything in their path.

So many people have been lost among the enormous dunes in the Taklimakan Desert in northern Asia, that it has been given the name of the "land of no return".

The Hottest of All

The Danakil Depression in the Danakil Desert in Ethiopia is one of the hottest place on Earth. The rocks there can get as hot as 160°C (320°F), hot enough to roast a chicken or fry a steak.

Did You Know?

Death Valley in California is one of the hottest places on Earth. It got its name in 1849 when prospectors, searching for gold, ran out of food and water. They died before they could escape the burning heat.

The Sahara Desert, which is almost as big as the USA, was not always a desert. Millions of years ago it was covered in ice.

The Atacama Desert in South America has less than $1\frac{1}{4}$ cm ($\frac{1}{2}$ in) of rain each year. In some parts of the desert, it has never rained. Others had no rain for 400 years.

The highest temperature ever recorded was 58°C (136°F) at Al Aziziyah, in Libya, on September 13th, 1922. And this was taken in the shade!

A tribe of fierce nomads, called the Afar, live in the Danakil. Some of them mine salt from Lake Karum and take it to market by camel train. The miners build their houses from blocks of salt.

Water Storers

Giant cacti, called saguaros, store water in their huge, thick stems. They are found in American deserts and can grow 18 m (60 ft) high, weigh as much as two elephants and live for over 250 years.

Desert People

The Bushmen of the Kalahari Desert store water in ostrich shells buried in the ground. They suck it out through hollow reeds. When there is little food about, bushmen survive on fat stored by their bodies.

The Bindibu aborigines who live in Australia are nomads. They roam the desert hunting for food and water. A hunter can kill game up to 75 m (250 ft) away with a boomerang or a spear.

Icy Cold Places

The Arctic and the Antarctic are the coldest places on Earth. Most of the ground is frozen solid and covered in ice and snow all year around. For six months in winter, the sun never rises and it is dark all the time.

In some places, the ice is more than 4 km ($2\frac{1}{2}$ miles) thick. If all the ice melted, the sea would rise so much that many of the world's major cities, such as London, Tokyo and New York, would be covered by water.

A Frozen Land

Greenland, which lies within the Arctic circle, is the second biggest island in the world. The temperature hardly ever rises above freezing and it snows every month of the year.

Many Inuits, or Eskimos, live in Greenland. Sometimes they fish for trout through holes cut in the frozen lakes and hunt seals as they come up for air at breathing holes in the ice.

Mountains of Ice

Sometimes huge pieces of ice break off glaciers and crash into the sea, forming icebergs. Only about $\frac{1}{10}$ of an iceberg floats above the water, the rest is hidden below.

Breaking the Ice

Ships, called icebreakers, are used to smash a way through the ice . Their specially shaped bows and hulls can break ice that is more than 7 m (23 ft) thick.

Behind the Screens

Inuits often hide behind shooting screens when they go hunting. The screens are made of white fabric so they can creep up on animals in the snow without being seen.

Did You Know?

In 1956, the biggest iceberg ever recorded was seen off the coast of Antarctica. It covered about 31,000 sq km (11,970 sq miles).

The Trans-Alaska Pipeline carries oil across three mountain ranges, about 300 rivers and streams and nearly 640 km (400 miles) of frozen land. The oil has to be heated inside the pipes to stop it from freezing.

In the Antarctic, Weddell seals spend most of their time under the water. They can dive down 600 m (2,000 ft) and hold their breath for at least 45 minutes.

The lowest ever temperature was recorded at Vostok Station, Antarctica, on July 21st, 1983. It was −89°C (−129°F).

The small village of Oymyakon in Siberia, USSR, is the coldest place in the world where people live all year around.

The Land of the Lapps

The people of Lapland are called Lapps. Today, most Lapps live and work in small towns. A few still keep huge herds of reindeer and wander from place to place where their reindeer can find food.

They live in tents made from reindeer skins, eat reindeer meat and drink their milk.

On Top of the World

A quarter of the Earth's land surface is covered by mountains over 1,000 m (3,280 ft) high. They may be single peaks or form part of a huge mountain range.

Mountains affect the world's climate, and where and how people live. As the temperature falls, by 2°C every 300 m (990 ft) up, it becomes more and more difficult for plants and animals to survive on mountain slopes.

High and Mighty

The Andes Mountains stretch 7,200 km (4,474 miles) along the west coast of South America. Many of the peaks are over 6,100 m (20,000 ft) high and some are active volcanoes.

Once a Coral Reef

Long ago, the Dolomites, in Italy, were coral reefs under the ocean. Over millions of years, the coral turned into a limestone rock known as dolomite.

Underwater Mountain

Mauna Kea, on the island of Hawaii, is the world's highest island peak. From its base, which rests on the bottom of the ocean, to its tip, it is 10,203 m (33,476 ft) high. That is 1,355 m (4,448 ft) more than Mount Everest, the highest mountain in the world.

Machu Picchu, an ancient Inca city, lies high up in the Andes Mountains. The Incas were South American Indians whose huge empire was destroyed in 1532.

The name Andes is thought to have come from the word "anti", an Indian name for copper. Copper, gold, silver, tin, iron and platinum are still found in these mountains.

The Andes are rising by about 10 cm (4 ins) every 100 years.

Mountains of Marble

The slopes of the Apuane Alps in Tuscany, Italy, contain marble which has been quarried there for over 2,000 years. Famous sculptors, such as Michelangelo and Henry Moore, used the marble for their work.

Carved by Ice

A glacier is a moving river of ice. As it grinds slowly down a mountain, it carves out a valley. Millions of years ago, the Matterhorn, in Switzerland, was carved by glaciers into its pointed shape.

Climbing to over 4,816 m (15,800 ft), the Andes Central Railway is the highest standard railway in the world.

Did You Know?

The world's longest mountain chain is almost completely underwater. The Mid-Atlantic Ridge stretches more than 16,000 km (9,942 miles) from the North Atlantic Ocean all the way to Antarctica.

Reaching a height of 8,848 m (29,028 ft), Mount Everest in the Himalayas is the tallest mountain above sea level.

Sometimes the thick ice of a glacier splits and makes deep cracks called crevasses. The ice in these crevasses can tell scientists about changes in the world's climate.

Desert Fortress

In the Sahara Desert, in Africa, the Ahaggar Mountains are the remains of extinct volcanoes.

Over millions of years, wind and frost have changed the peaks into strange shapes.

Mountains of Fire

There are more than 600 active volcanoes on Earth. Others are dormant and may not erupt for hundreds of years. When a volcano explodes, boiling-hot liquid rock, called lava, bubbles up and bursts out of its crater, releasing clouds of ash, cinders and gases.

An explosion can cause terrible damage. Thousands of people may die from the choking gases, poisonous fog and avalanches of thick mud. Many people risk the danger of living near an active volcano because the ash from it makes the soil very fertile.

Red-hot Rivers

In 1977, when the huge volcano of Nyiragongo, in Zaire, erupted, a red-hot river of lava gushed over the land at 96 km (60 miles) an hour. It poured out of a crack 20 km (12 miles) long.

A Sea of Mud

When Nevado del Ruiz, in Columbia, erupted in 1985, the burning heat melted the ice and snow on the peak. Mud poured down the Lagunillas River into Armero killing 23,000 people.

A New Island

Sometimes volcanoes erupt under the sea. In 1963, a new island, called Surtsey, appeared when a volcano erupted off the coast of Iceland. That night the new island was 11 m (36 ft) high.

Blowing its Top Off

About 400 m (1,300 ft) were blasted off the top of Mount St. Helens, in America, when it exploded in 1980. Boiling-hot gases, steam, ash, huge boulders and blocks of ice rocketed into the sky. Over two million birds, animals and fish died.

A Humming Volcano

Mount Etna, on the island of Sicily, makes strange singing noises. Liquid rock roars and hums as it moves inside the volcano. Scientists can tell when it may next erupt by listening to the noises.

Did You Know?

Suddenly, on August 27th, 1883, the volcano on the island of Krakatoa, near Java, exploded. It blew up the whole island sending ash, fire, gas and rocks 80 km (50 miles) up in the air.

The explosion was the loudest noise ever recorded and was heard in Australia, 4,800 km (2,986 miles) away.

It burst the eardrums of half the crew on a ship about 40 km (25 miles) away.

The shock was felt over 14,500 km (9,010 miles) away in the USA.

The wind blew clouds of volcanic dust around the world, causing brilliant sunsets.

A tsunami, 30 m (98 ft) high, rolled in from the ocean drowning 36,000 people.

Jet Lag!

In 1982, a jet flew into a thick cloud. One by one, the engines stopped and the jet dropped 7,400 m (24,278 ft) before the pilot was able to restart them. The cloud was volcanic dust.

Mount Etna is the highest active volcano in Europe. It has erupted 260 times since its first recorded eruption in 700 BC. In 1669, 20,000 people died in a huge explosion.

When the Earth Moves

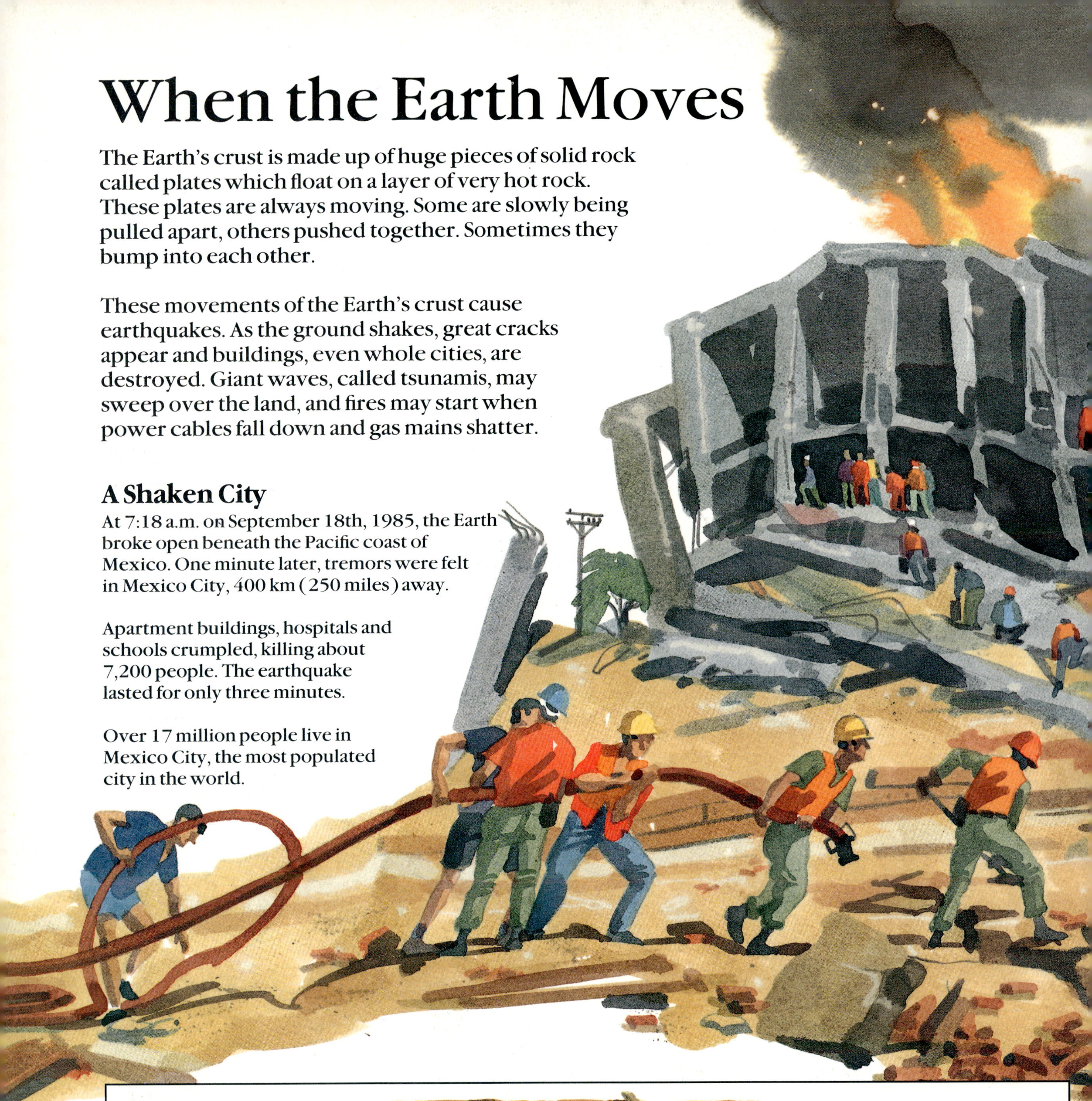

The Earth's crust is made up of huge pieces of solid rock called plates which float on a layer of very hot rock. These plates are always moving. Some are slowly being pulled apart, others pushed together. Sometimes they bump into each other.

These movements of the Earth's crust cause earthquakes. As the ground shakes, great cracks appear and buildings, even whole cities, are destroyed. Giant waves, called tsunamis, may sweep over the land, and fires may start when power cables fall down and gas mains shatter.

A Shaken City

At 7:18 a.m. on September 18th, 1985, the Earth broke open beneath the Pacific coast of Mexico. One minute later, tremors were felt in Mexico City, 400 km (250 miles) away.

Apartment buildings, hospitals and schools crumpled, killing about 7,200 people. The earthquake lasted for only three minutes.

Over 17 million people live in Mexico City, the most populated city in the world.

Snake Warning

Some animals behave strangely before an earthquake. In 1975, in China, hibernating snakes suddenly left their holes. Thousands of people were saved because of the snakes' warning.

The maternity centre at the National Medical Centre was completely crushed. Many days later several new-born babies were dug out alive. They had survived because of the fat and water in their bodies.

Four hundred families lived in the huge building of Nuevo Leon. Floor by floor, the 14 storeys collapsed on top of each other, crushing everything and everyone in between them.

Quake-proof?

In San Francisco, USA, special earthquake-proof buildings have been built. They are built on concrete rafts that "float" on the ground when the Earth moves.

Weighty Water

As the Kariba Dam on the Zambezi River began to fill with water, the area was shaken by more than 2,000 strong earth tremors. The water was forcing the underground rocks to move.

Rocking and Rolling

On Good Friday, 1964, the ground underneath Anchorage, Alaska, began to roll like a huge wave, setting off a giant landslide. Houses slid down deep cracks which opened up the ground.

Did You Know?

Earthquakes are measured on the Richter Scale. Each number on the Scale represents an earthquake 10 times stronger than the one below. Quakes measuring seven or more are very strong. The one in Mexico City measured 8.1.

There are about one million earthquakes every year but most of them are tiny, or happen under the sea and do no harm.

During an earthquake in Assam, India, in August 1950, large boulders exploded like bombs and deep ravines appeared in the ground. About 2,000 people died in the earthquake.

Fresh Waters of the World

About 70% of the Earth's surface is covered in water. Only 3% is fresh, the rest is salty. Most of the Earth's fresh water is frozen, in glaciers or in ice sheets. Less than 1% is found in lakes, rivers and under the ground.

People cannot live without fresh water. As well as drinking it, we use it to grow crops. Many lakes and rivers are so polluted by chemical waste, sewage and rubbish that the water is poisonous and can cause diseases.

The Mighty Amazon

The Amazon is the biggest river on the Earth. From its mouth pours $^{1}/_{5}$ of all the river water in the world.

It starts as a tiny stream over 5,240 m (17,200 ft) up in the Andes Mountains and flows down into the Atlantic, almost 6,437 km (4,000 miles) away.

More than 200 rivers, with a total length of 80,000 km (49,710 miles), join the Amazon.

Strange animals, called capybaras, the biggest rodents in the world, live in large family groups on the banks of the river.

Many kinds of fish including fierce, flesh-eating piranhas and huge cat-fish live in the Amazon.

Sometimes a huge wall of water, called a bore, forms at the mouth of the river. It rolls back up the river like a giant wave.

High Lake

High up in the Andes Mountains, in South America, about 3,812 m (12,507 ft) above sea level, lies Lake Titicaca. One of the world's highest lakes, it is deep enough for large ships to cross.

Angel's Gold

The Angel Falls plunge 979 m (3,212 ft) down the side of Devil Mountain in Venezuela. They were first discovered in 1935 by an American pilot, called Jimmy Angel, when he flew over them searching for gold.

There are many islands in the lake. The Uros Indians, who live on the islands and around the shores, build their houses and fishing boats from reeds, called totoras, which grow in the water.

The Holy River

Hindus believe that the River Ganges, in India, is sacred. Temples line the river banks and steps, called ghats, lead down to the water. Thousands of Hindus make pilgrimages to bathe in the holy river every year.

Flamingo Heaven

The shallow waters of Lake Natron, in Tanzania, are full of soda and minerals. Thousands of flamingoes flock to the lake to feed on tiny plants, called algae, which are the only things that can survive in the lake.

Did You Know?

In 1859, a Frenchman called Charles Blondin, walked along a tightrope 48 m (160 ft) above the Niagara Falls. The rope was over 330 m (1,082 ft) long.

About 20,000 bathtubs of water crash over the Niagara Falls every second.

The Cuyahoga River in Ohio, USA, is so polluted that it is classed as a fire hazard. In 1969, it actually caught fire.

When Mount Mazama in Oregon, USA, erupted 7,000 years ago, the explosion was so violent that the top of the mountain caved in. It left a huge round crater which gradually filled with water and formed a lake called Crater Lake.

Steamy Forests

Over half the world's tropical rainforests have been cut down and destroyed. What remains, contains 50% of all the world's species of animals and plants, as well as 80% of all types of insects.

Forests, the size of Britain, are cut down every year to clear land for crops and cattle, to build roads and to provide wood for fuel and timber. They are cleared for mines and flooded when dams are built for hydro-electric schemes.

Teeming with Life

A few square kilometres of rainforest may contain over 700 types of trees, 1,000 types of flowering plants, nearly 450 species of birds, 100 different animals, and over 100 types of butterflies and moths.

The rainforests have several layers. The tallest trees may grow as high as 61 m (200 ft).

Climbing plants, called lianas, loop themselves from tree to tree like huge ropes. They can be 60 cm (2 ft) thick and over 150 m (495 ft) long.

Forest of Stone

In one of the strangest forests in the world all the trees are stone. This petrified forest, in Arizona, USA, formed millions of years ago, when the trees were buried under sand and mud.

Lazy Bones

Strange animals, called sloths, spend their lives hanging upside-down from trees in the rainforest. They move so slowly and are so dirty that tiny plants grow on their long fur, making them look green.

The trees and plants produce oxygen and remove gases, such as carbon dioxide from the air.

People who live in rainforests make medicines from many of the plants. At least 1,400 forest plants are being used to treat cancer, malaria, ulcers and many other illnesses.

Did You Know?

Some people who live in the rainforests eat the larvae of caterpillars and beetles which are full of protein.

About 40 years ago, scientists found a forest growing deep inside the Sarisarinama Plateau in Venezuela. The trees were growing among huge boulders 314 m (1,036 ft) down.

The loneliest tree in the world is a Norwegian spruce growing on Campbell Island, Antarctica. The nearest tree to it is over 145 km (90 miles) away.

Most of the rainforest trees are evergreen. Their dark, tough leaves have special tips to let the heavy rain drip off.

Smelly Flowers

Rafflesias grow in forests of South-East Asia. They are the biggest flowers in the world and can measure 1 m (3.3 ft) across. They smell like rotting meat to attract the flies that pollinate them.

A Stinky Bird

The hoatzin lives in the South American forests. Called the "stinkbird", it smells like cow manure because the leaves it eats stay in its stomach for almost two days to digest.

Stormy Weather

The most common storms are thunderstorms. About 45,000 happen every day around the world. Huge clouds form in the sky and strong winds begin to blow, bringing rain and hail, thunder and lightning.

The most violent winds are caused by tornadoes and hurricanes. Reaching speeds of over 320 km/h (200 mph), these twisting, whirling storms tear across the land and sea destroying everything in their paths.

Whirling Winds

Tornadoes are spinning funnel-shaped storms which reach down from the base of huge clouds. When the funnel touches the ground, the powerful, whirling winds inside it suck up anything in their path, even people.

They can twist the tops off trees, overturn railway carriages and carry cars, people and big animals hundreds of feet. At the centre of the storm, winds may reach at least 372 km/h (231 mph).

As Big as Tennis Balls

On July 12th, 1984, a giant hailstorm hit Munich, Germany. Rock-hard hailstones, as big as tennis balls, destroyed crops, took roofs off houses and damaged thousands of cars.

Huge Hurricanes

Hurricanes form over warm oceans. They can be 480 km (300 miles) wide with raging winds which may reach 240 km/h (150 mph). The winds spin around the "eye" which is a calm area in the centre of the storm.

Along coastlines, hurricanes can whip up huge waves, called a storm surge. Heavy rains swamp the land, often causing terrible floods.

Did You Know?

When sunlight hits raindrops or a spray of water, two rainbows may appear at the same time. The colours of the second rainbow are always the other way around from those of the first.

The strangest kind of lightning is ball lightning. About the size of a grapefruit, it looks like a glowing red ball. People have reported seeing it floating along the ground, inside houses, and even inside aeroplanes.

In 1939, a shower of frogs fell on Kent, in England. They had been sucked up from nearby ponds and streams by a strong wind.

A killer tornado tore across Missouri, Illinois and Indiana, USA, in 1925. It lasted for over three hours, uprooted huge oak trees, hurled cars over rooftops and killed about 700 people.

Hurricanes turn clockwise south of the Equator and anti-clockwise to the north.

A Clap of Thunder

Before a thunderstorm, electricity builds up in the clouds until the lightning flashes. Because light travels faster than sound, we see the lightning before we hear the thunder.

Going Places

Until about 5,000 years ago there were no roads and people travelled on foot or in boats. Today, people can travel by supersonic jet at speeds of over 2,300 km/h (1,430 mph), by high-speed trains that can go as fast as 270 km/h (168 mph), and even by hot-air balloons. One day it may be quite common to travel into Space by Space Shuttles.

Up, Up and Away

At the 1987 International Balloon Festival in Bristol, in Britain, 128 hot-air balloons took off within the space of one hour.

Big Carriers

In Thailand and Nepal, people still work and travel about on elephants. There are special training schools where elephants learn how to lift and carry people, and drag heavy logs.

Underwater Journey

A special submarine called Atlantis I, in the Grand Cayman Islands, takes visitors about 45 m (150 ft) down into the sea. Floodlights on the submarine show up the bright corals and sea creatures.

The biggest hot-air balloon ever flown was "Virgin Atlantic Flyer". In July 1987, it flew 4,974 km (3,047 miles) across the Atlantic Ocean. It took just under 32 hours, reaching speeds of over 209 km/h (130 mph).

Swarms of Bicycles

In China, only one person in 500 has a car. Most people travel by bicycle. Some bicycles have side-cars to carry babies, or rear-cars for older people. There are special places to park bicycles.

Did You Know?

The worst traffic jam happened in 1980, in France. It was 176 km (109 miles) long.

In 1988, 144 parachutists joined together for almost 9 seconds in the skies above Quincy, Illinois, USA. It was the world's biggest free-fall parachute jump.

Valery Ryumin, a Russian astronaut, has travelled around the world 5,750 times, covering over 240 million km (150 million miles). He has spent almost a year in space.

In 1930, two men drove their car in reverse all the way from New York to Los Angeles, about 5,375 km (3,340 miles), without once stopping the engine.

Under the Ground

The Moscow Metro is the busiest in the world. More than 6.5 million people travel on it every day. The stations are decorated with marble columns, chandeliers and paintings.

Natural Wonders

The world is full of amazing and spectacular natural structures of all shapes and sizes. There are deep caves and canyons, gigantic rocks, marble mountains, extraordinary trees and even forests that have turned to stone.

The forces which cause these natural wonders never stop, changing the Earth's landscape all the time. Some are caused by the movement of the Earth, earthquakes and volcanoes, others by wind, water or ice.

A Grand Canyon

The Grand Canyon in Arizona, USA, is one of the most spectacular sights in the world. Over millions of years, the Colorado River has flowed through and worn down the layers of rock to form a vast and deep gorge.

The Canyon is 446 km (277 miles) long, up to 29 km (18 miles) wide and over 1.6 km (1 mile) deep.

Reaching for the Sky

Devil's Tower, in Wyoming, USA, looms 265 m (869 ft) above the Belle Fourche River. An Indian legend claims that a giant bear made the huge marks down the sides as it tried to reach people stranded on the top.

Wear and Tear

Landscape Arch in Utah, USA, is 88 m (291 ft) long, the longest natural arch in the world. Natural arches form when water seeps into cracks in the rock and widens them. Ice, rain and wind wear away the softest parts of the rock until only an arch remains.

The oldest layers of rock at the bottom of the canyon are over 2,000 million years old.

It is up to 14°C (25°F) hotter and it rains less at the bottom of the Canyon than at the top.

Huge dinosaur footprints have been found in the sandstone.

White Rocks

This mountainside in Turkey looks like a frozen waterfall. Hot springs full of mineral salts bubble out of the mountain. As the water evaporates, it leaves behind minerals which form these strange rocks.

Did You Know?

Ayers Rock, in central Australia, is one of the biggest rocks in the world. Over 335 m (1,100 ft) high, it can be seen from almost 100 km (60 miles) away. Long ago, it became a sacred place for the Aborigines who call it "Uluru". Many of the cave walls are covered with paintings which tell stories of their lives.

The largest cave in the world is the Sarawak Chamber, in the Gunung Mulu National Park, Sarawak, Malaysia. It is so big that about 15,000 cars could be parked inside it.

The longest stalactite in the world is 59 m (195 ft) from top to bottom. This is about as tall as 40 men standing on each other's shoulders.

Balancing Rocks

These strange looking rocks in Australia are called "The Devil's Marbles". Made of granite, they were probably once a part of a huge solid block which broke up and was gradually rounded into balls by the wind and the rain.

Forest Giants

Giant redwood trees, which only grow in California, USA, are the tallest trees in the world. Over 75 m (250 ft) tall, they can have trunks 11 m (36 ft) thick and weigh over 2,000 tonnes.

Deep Down

Formed over millions of years, huge stalactites hang from the ceiling in the Carlsbad Caverns in New Mexico, USA. Stalagmites rise from the floor like pillars.

There are sometimes over 100,000 bats living together in the caves. In late October, around 5,000 bats a minute, leave the caves to fly off to warmer places.

Did You Know?

A Spout of Water
On May 16th, 1898, a waterspout, about 1,530 m (5,049 ft) high, was spotted off the coast of New South Wales, Australia. Spinning and twisting above the sea, it is the biggest ever recorded.

A Sinking Feeling
Every year, Mexico City sinks about 17.5 cm (7 ins). This is because it is built over an underground lake. Slowly, the water is being used up by the 17 million people who live in the crowded city.

The Longest Rivers

Nile	Africa	6,671 km (4,145 miles)
Amazon	South America	6,280 km (3,902 miles)
Mississippi	USA	6,019 km (3,741 miles)
Yangtze	China	5,520 km (3,430 miles)
Yellow	China	5,464 km (3,395 miles)

Deadly Hot
Without clothes, food, water and shade it is impossible to survive one day in a hot desert. A person would sweat so much that he would lose about 8 litres (15 pints) of water from his body in just ten hours.

Set on Fire
More than 99,000 people died after a huge earthquake struck Tokyo, Japan, in 1923. Fire spread through the city when power cables fell down, gas mains shattered and cooking stoves set houses alight.

The Biggest Lakes

Caspian Sea	Iran/USSR	371,000 sq km (143,243 sq miles)
Superior	Canada/USA	82,414 sq km (31,820 sq miles)
Victoria	Africa	68,100 sq km (26,293 sq miles)
Aral Sea	USSR	66,500 sq km (25,676 sq miles)
Huron	Canada/USA	59,596 sq km (23,010 sq miles)
Michigan	USA	58,016 sq km (22,400 sq miles)

Ancient Earthquakes
The Chinese used to believe that the Earth was supported by an ox. They thought that earthquakes were caused by the ox shifting its weight from one shoulder to the other. The Japanese believed that earthquakes were caused by the sudden wriggling of a giant catfish, buried in the mud underneath the Earth.

Dachshund Sausages
Frankfurter sausages are named after the town of Frankfurt, in Germany. In 1900, an American selling frankfurters started calling them "dachshund sausages" because they looked so like the dogs. Soon they became known as "hot dogs".

The Largest Populations

China	1,072,300,000
India	796,600,000
USSR	286,717,000
USA	245,600,000

Strike it Lucky
An American, called Roy C. Sullivan, is the only man to survive being struck by lightning seven times. The first time was in 1942 and the last in 1977. He has had his hair and his eyebrows set on fire, lost a toenail and suffered several burns.

The Biggest Countries

USSR	22,274,900 sq km	(8,600,347 sq miles)
Canada	9,922,330 sq km	(3,831,015 sq miles)
China	9,560,939 sq km	(3,691,482 sq miles)
USA	9,528,318 sq km	(3,678,887 sq miles)
Brazil	8,511,965 sq km	(3,286,472 sq miles)
Australia	7,686,850 sq km	(2,967,895 sq miles)

The Smallest Countries

Vatican City	0.4 sq km	(0.2 sq miles)
Monaco	1.5 sq km	(0.6 sq miles)
Nauru Island (Pacific Ocean)	21 sq km	(8.2 sq miles)
Tuvalu Island (Pacific Ocean)	26 sq km	(10 sq miles)

Above it All

The tallest building in the world is the Sears Tower in Chicago, USA. It is 110 storeys high, has 103 lifts, 16,000 windows and 18 escalators. Over 25,000 people enter it every day.

A Tight Squeeze

About 11,600,000 people live in Tokyo, the capital of Japan. The trains that carry people to work are always crammed full. Special people, called pushers, work at some stations. Their job is to push as many people as possible into each train.

In Deep Water

Lake Baikal, in USSR, is the world's deepest lake. It also contains about $^1/_5$ of all the Earth's unfrozen fresh water. At its deepest point, 1,940 m (6,365 ft), you could stand almost $4^1/_2$ Empire State Buildings on top of each other under the water.

When Lightning Strikes

It is quite safe to be in a car during a fierce thunderstorm. If lightning strikes, the rubber tyres conduct the electricity away from the car into the earth.

Every Second Counts

In 1988, about 766,000 aeroplanes took off and landed at Hartsfield Airport in Atlanta, Georgia, USA. That means there was a take-off or landing about every 40 seconds.

Fit for a King

The Pyramids of Giza, on the west bank of the River Nile, were built over 4,000 years ago as tombs for Egyptian kings. More than 2 million stone blocks were used to build the Great Pyramid. Each one weighed 2.3 tonnes, about the same as 19 baby elephants.

The Highest Mountains

(By Continent)

Mount Everest	China/Nepal	8,848 m	(29,028 ft)
Aconcagua	Argentina	6,960 m	(22,835 ft)
McKinley	Alaska	6,194 m	(20,322 ft)
Kilimanjaro	Tanzania	5,895 m	(19,340 ft)
Mount Elbrus	USSR	5,642 m	(18,510 ft)
Vinson Massif	Antarctica	5,140 m	(16,864 ft)

A Wooden General

A giant sequoia tree in California, known as "General Sherman", is the biggest living thing on Earth. It is over 83 m (274 ft) tall and about 25 m (82.5 ft) around the trunk. About 3,000 years old, it is thought to contain enough wood to make 4 billion toothpicks.

Index

Edited and designed by Mitchell Beazley International Ltd.
Artists' House, 14–15 Manette Street, London W1V 5LB.

British Library Cataloguing in Publication Data for this book are available from the British Library.

Typeset in Garamond ITC Book by Tradespools Ltd, Frome.
Reproduction by Mandarin Offset, Hong Kong.
Produced by Mandarin Offset.
Printed and bound in Hong Kong.